MW01327865

Animals with Bite

Hippopotamus

by Julie Murray

Dash!

LEVELED READERS

1

An Imprint of Abdo Zoom • abdobooks.com

Level 1 – Beginning
Short and simple sentences with familiar words or patterns for children who are beginning to understand how letters and sounds go together.

Level 2 – Emerging
Longer words and sentences with more complex language patterns for readers who are practicing common words and letter sounds.

Level 3 – Transitional
More developed language and vocabulary for readers who are becoming more independent.

abdobooks.com

Published by Abdo Zoom, a division of ABDO, PO Box 398166, Minneapolis, Minnesota 55439.

Printed in the United States of America, North Mankato, Minnesota.
102020
012021

Photo Credits: iStock, National Geographic Image Collection, Shutterstock
Production Contributors: Kenny Abdo, Jennie Forsberg, Grace Hansen, John Hansen
Design Contributors: Dorothy Toth, Neil Klinepier

Library of Congress Control Number: 2020910902

Publisher's Cataloging in Publication Data

Names: Murray, Julie, author.
Title: Hippopotamus / by Julie Murray
Description: Minneapolis, Minnesota : Abdo Zoom, 2021 | Series: Animals with bite | Includes online resources and index.
Identifiers: ISBN 9781098222994 (lib. bdg.) | ISBN 9781098223694 (ebook) | ISBN 9781098224042 (Read-to-Me ebook)
Subjects: LCSH: Hippopotamus--Juvenile literature. | Hippopotamus--Behavior--Juvenile literature. | Aquatic mammals--Juvenile literature. | Bites and stings--Juvenile literature. | Predatory animals--Juvenile literature.
Classification: DDC 591.53--dc23

Table of Contents

Hippopotamus 4

More Facts 22

Glossary 23

Index 24

Online Resources 24

Hippopotamus

Hippos live in Africa. They spend most of the day in water. This keeps them cool.

At night, hippos walk to find food. They eat grass, leaves, and fruit.

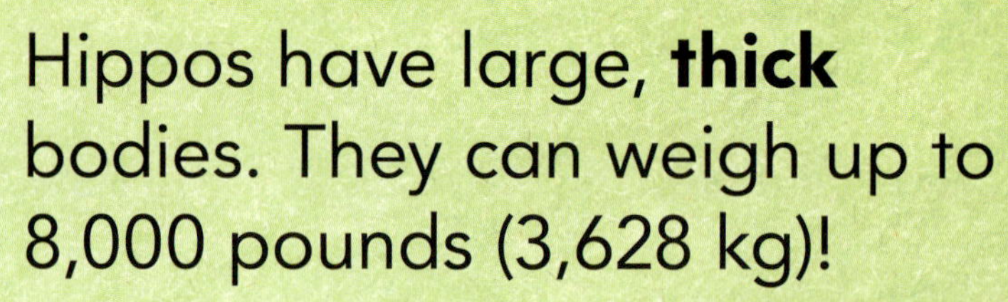

Hippos have large, **thick** bodies. They can weigh up to 8,000 pounds (3,628 kg)!

They have short legs. They stand 5 feet (1.5 m) tall.

Their nose, eyes, and ears sit high on their heads. This allows them to keep most of their bodies below water. But they can still breathe, see, and hear.

They are gray-brown in color. Their skin is very **thick**. This helps keep them safe from **predators**.

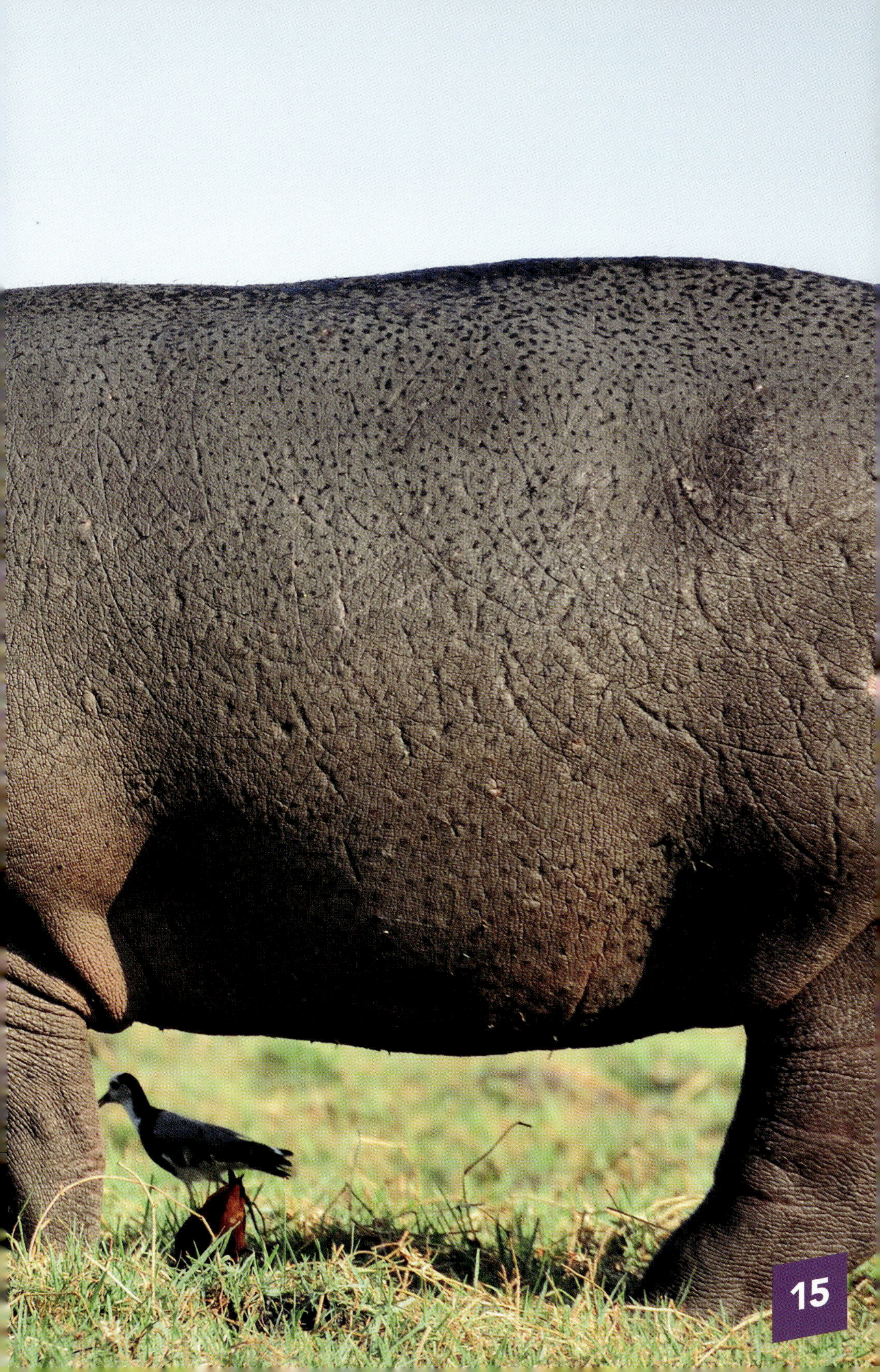

Hippos have huge mouths. Their mouths can be 5 feet (1.5 m) long when open!

Hippos have large teeth. They have **tusks** that can be 1.5 feet (0.46 m) long!

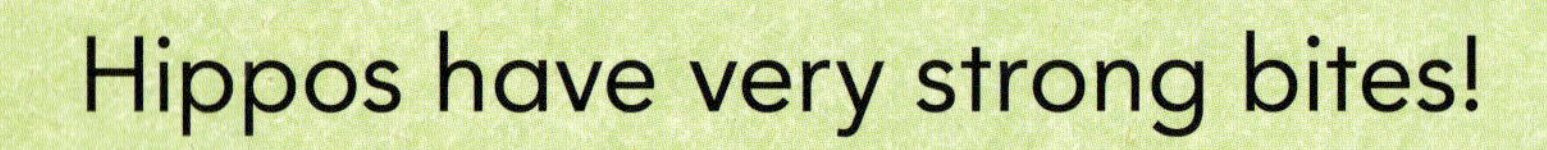

Hippos have very strong bites!

More Facts

- Hippos can't swim. Instead, they bounce "walk" underwater.
- They can hold their breath for five minutes.
- Hippos live in groups. The group has 5 to 30 members.

Glossary

predator – an animal that hunts another animal for food.

thick – relatively large in measurement from one side to the other; not thin.

tusk – a long, large pointed tooth that sticks out from the mouth of some animals. Tusks grow in pairs and may be used to fight or find food.

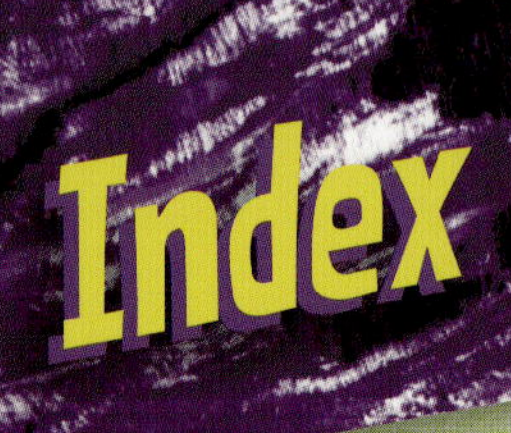

Africa 4

bite 20

body 8, 12

color 14

food 6

head 12

height 11

legs 11

mouth 17

skin 14

teeth 18

tusks 18

water 4, 12

weight 8

Online Resources

Booklinks
NONFICTION NETWORK
FREE! ONLINE NONFICTION RESOURCES

To learn more about hippopotamuses, please visit **abdobooklinks.com** or scan this QR code. These links are routinely monitored and updated to provide the most current information available.